Maria Edvânia Neves Barros
Guillermo Gamarra Rojas
Alfredo M. de Sousa

The garden of the Itaoca Women's Movement

Maria Edvânia Neves Barros
Guillermo Gamarra Rojas
Alfredo M. de Sousa

The garden of the Itaoca Women's Movement

Contribution to health and social development

ScienciaScripts

Imprint

Cover image: www.ingimage.com

This book is a translation from the original published under ISBN 978-3-330-76540-5.

Publisher:
Sciencia Scripts
is a trademark of
Dodo Books Indian Ocean Ltd. and OmniScriptum S.R.L publishing group

120 High Road, East Finchley, London, N2 9ED, United Kingdom
Str. Armeneasca 28/1, office 1, Chisinau MD-2012, Republic of Moldova, Europe
Managing Directors: Ieva Konstantinova, Victoria Ursu
info@omniscriptum.com

Printed at: see last page
ISBN: 978-620-8-51942-1

PREFACE

This book arose from the idea of presenting society with a collection of information on the cultivation, use and indications of the main Brazilian medicinal plants.

The use of medicinal plants is intertwined with the history of the human species, the former being as old as the latter. Medicinal plants are still widely used in today's Brazilian society and around the world, and are commercialised at open-air fairs, popular markets and home fairs.

The use of medicinal plants is of great importance in many communities, and in many of them it is the only therapeutic resource available.

Brazil has the greatest diversity of medicinal plants in the world, so large quantities of the world's fruits, leaves, teas and herbs are concentrated in our country. All this richness, however, needs to be further studied so that it can be used in the safest and most satisfactory way.

The book's basic purpose is to describe a farm that produces medicinal herbs and whose main source of income is this activity. The system is set up on an agro-ecological basis and, to a certain extent, shows the economic potential of this market.

The final product of this work is a milestone in the dissemination of the use of this ancient tradition, given the relevance of the subject and the scarcity of technical literature on the subject.

The authors

SUMMARY

The use of medicinal plants is an ancient practice that contributes to quality of life and the valorisation of knowledge passed down from generation to generation. With this in mind, the aim of this study was to understand how the cultivation, use and commercialisation of medicinal plants contributes to the perpetuation of popular knowledge expressed through the experience and work of the Women's Movement of Itaoca. The research carried out with the Women's Movement of Itaoca is both qualitative and quantitative in nature. It relies on the direct collaboration of the members who make up this group and was carried out in the physical facilities of the Valdivino de Carvalho Health Centre, located in Parangaba-Fortaleza Ceará. The idea for Farmácia Viva was thought up jointly by a group of women who, with the help of the Joazeiro Institute, developed a community garden project in the neighbourhood where they would start working with medicinal plants. The amount generated during the year by the medicinal plant garden configures its productive potential, which is represented by the annual gross product (GP) (total gross production of products handled + total gross production of seedlings produced). The study made it possible to visualise the organisation developed by the group to manage the activity, as well as helping to understand the socio-economic situation. The subjects' perception of the activity shows the degree of knowledge and affinity they have for spreading the use and reviving the tradition of medicinal plants.

Keywords: Urban agriculture. Medicinal Plants. Popular knowledge.

Summary

CONCEPT AND A BRIEF HISTORY

The use of medicinal plants in an attempt to combat illness and disease is as old as the history of mankind. The habit of resorting to the healing virtues of certain plant species has been present in humanity since the earliest civilisations we can remember.

The realisation that civilisations on different continents have managed to develop research or the search to identify the therapeutic properties of different plants even before their dietary properties, gives us the idea that the search to combat one of man's greatest concerns, which is the search to solve problems such as muscle pain, malaise, etc., has always been present in human history. It's interesting to realise that this ancient tradition has survived so many technological changes, changes in consumer habits, and still remains active and widespread in the most different Brazilian and world cultures.

There are various definitions of medicinal plants in the literature, all of which obviously include the therapeutic aspect of these plant species.

According to the National Health Surveillance Agency (ANVISA), medicinal plants are those capable of relieving or curing illnesses and have a tradition of use as a remedy in a population or community. According to Resolution RDC no. 10/2010 Art. 3 [...] Subsection XII, medicinal plants are plant species, cultivated or not, used for therapeutic purposes.

CHAPTER 1

INTRODUCTION

The use of medicinal plants is an ancient practice that contributes to quality of life and the valorisation of knowledge passed down from generation to generation. This practice goes hand in hand with the need for low-cost cures, through tea, licks, infusions, compresses and other preparations that are not only low-cost but also easy to prepare. The difficulties faced by public or private health centres in receiving or buying industrial medicines make it necessary to look for other easily accessible remedies.

According to Maciel et al. (2002), knowledge of medicinal plants often symbolises the only therapeutic resource for many communities and ethnic groups. The use of plants to treat and cure illnesses is as old as the human race. Even today, in the poorest regions of the country and even in large Brazilian cities, medicinal plants are sold in open-air markets, popular markets and found in backyards.

When we talk about cultivating medicinal plants, we are conserving biodiversity, human health, food, the economy, the recovery of popular knowledge, organisation, social participation, gender and generation (AZEVEDO and MOURA 2010).

Popular observations on the use and efficacy of medicinal plants make an important contribution to publicising the therapeutic virtues of plants, which are often prescribed for the medicinal effects they produce, despite the fact that their chemical constituents are not known. In this way, users of medicinal plants all over the world keep the practice of consuming natural remedies based on these plants in vogue, making valid therapeutic information that has been accumulated over centuries. Indirectly, this type of medicinal culture arouses the interest of researchers in studies involving multidisciplinary areas (MACIEL et a.l 2002).

Faced with the difficulties encountered at health centres, many people prefer to use medicinal plants that are popularly known. But is their use scientifically proven?

Therefore, care should be taken to only use plants that have been validated as medicinal.

CHAPTER 2

OBJECTIVES

2.1 General objective

To understand how the cultivation, use and commercialisation of medicinal plants contributes to the perpetuation of popular knowledge, expressed through the experience and work of the Women's Movement of Itaoca.

2.2 Specific objective

1. To survey and analyse the group's empirical knowledge of medicinal plants;
2. To identify the activities carried out and to ascertain the social relations that exist within the Itaoca Women's Movement;
3. To investigate the economic impacts of the production, processing and commercialisation of medicinal plants for the Itaoca Women's Movement.

CHAPTER 3

METHODOLOGY

3.1 Subjects and nature of the research

The research had the direct collaboration of the Itaoca Women's Movement, made up of five people. It is both qualitative and quantitative in nature, tending above all towards a qualitative emphasis. According to Appolinário (2009), research is considered qualitative when it involves collecting data from the researcher's social interactions with the phenomenon being researched. The research used the following approaches: analysis of agroecosystems (MASERA et al., 2000) and participant observation (VERDEJO, 2006) through periodic monitoring over the course of a year, with participation in the activities carried out in the garden.

3.2 Study area

The study was carried out in the physical facilities of the Valdivino de Carvalho Health Centre, located in Parangaba, Fortaleza, state of Ceará (Figure 1), a site provided by the Fortaleza city hall for the development of the activities of the Women's Movement of Itaoca.

Figure 1: Location of the Medicinal Plants Garden in Parangaba, Fortaleza, CE.
Source: Research data.

3.3 Research stages and instruments used

Figure 2 shows the flowchart of the research stages with details of the data collection instruments and information used.

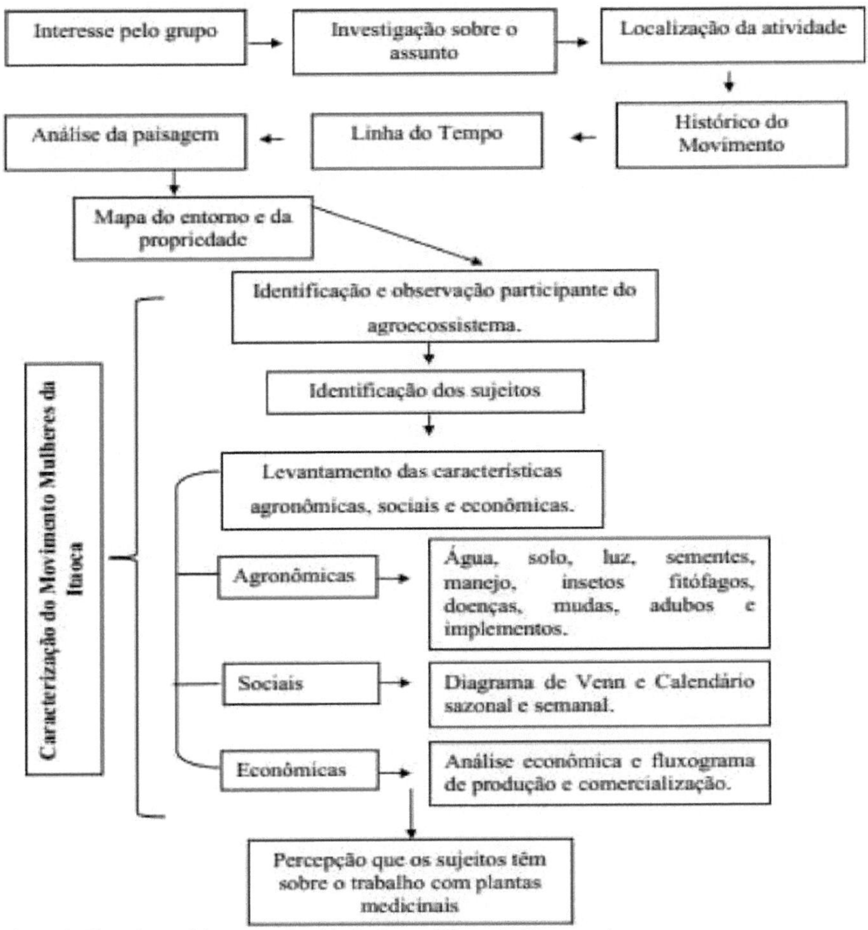

Figure 2: Flowchart of the research with the Itaoca Women's Movement.
Source: Research data.

2.2.1 Landscape analysis

There was a walk around the site and a round table discussion with the members, where they shared their thoughts on the work carried out and the benefits it had generated. From this observation, the data was used to build a map of the site.

2.2.2 Participatory rural diagnosis

According to Verdejo (2006), Participatory Rural Diagnosis (PRD) is a set of techniques and tools that allows communities to make their own diagnosis and from there begin to self-manage their planning and development. Participants will thus be able to share experiences and analyse their knowledge in order to improve their planning and action skills. This methodology aims to develop research processes based on the conditions and possibilities of the participants, drawing on their own concepts and explanatory criteria. Instead of confronting people with a list of previously formulated questions, the idea is for the participants themselves to analyse their situation and value different options for improving it. The aim is not just to collect data from the participants, but to get them to start a process of self-reflection on their own problems and the possibilities for solving them. Originally designed for rural areas, many of the DRP techniques can also be used in urban communities, as will be the case with using this methodology to better understand the reality of the Women's Movement of Itaoca, which works with medicinal plants in the city.

The main aim of DRP is to support community self-determination through participation and thus foster sustainable development.

According to Verdejo (2006), the instruments used in this research were:

History: Identifies facts, experiences and changes that have had a decisive influence on the development of the agro-ecosystem and the use of natural resources.

- **Map:** Encourages farmers to draw a map of their environment and establish functional relationships between the various components of the agro-ecosystem.
- **Timeline:** Used to visualise changes over time
- **Weekly and seasonal calendar:** Allow time-related aspects to be analysed. They can be applied to long processes or to a shorter period.
- **Venn diagram:** Identifies the organised groups in the community and the relationships they have with each other and with other local and regional institutions outside the community. According to Verdejo (2006), the Venn diagram identifies the community's organised groups and the relationships they have with each other and with other local and regional institutions outside the

community. Its purpose is to highlight the relationships established between community members and institutions in order to recognise the importance of these factors in decision-making processes and community development. By using this tool, you can see the relationship that the institution has with a particular partnership, whether the degree of affinity is close or not.

- **Production and commercialisation flowchart: This** shows all the economic flows of an organisation. Verdejo (2006) states that the production flow shows all the steps in the production of a given product, with the aim of analysing production in detail in order to improve it, while the marketing flow shows commercial flows in their entirety, allowing for an analysis of efficiency, weaknesses and commercial potential.

- Timeline

Identifying facts, experiences and changes that have had a decisive influence on the development of the agro-ecosystem and the use of natural resources. It allows a variety of subjects to be placed in a historical context, helps to "break the ice", and serves as a reference for planning the next stages of the DRP to be built with the family (VERDEJO, 2006). The Itaoca Women's Movement was brought together and the objectives of the tool were explained. Paper 40, cards, adhesive tape and coloured brushes were used for this activity.

- Venn diagram

To make the Venn diagram, the Itaoca Women's Movement was brought together and the objectives of the tool were explained. The following materials were used: a large piece of paper, cards, cardboard, paintbrushes. A circle was drawn in the centre of the paper to represent the Movement. Around it, petals were drawn close to the centre, identifying the institutions most present. The sepals were then made further away from the centre, representing the institutions that are less active in the movement. At the end of the activity, a reflection was made in order to identify the most active partnerships (those closest to the centre) and the least active (those furthest from the centre).

- **Participant observation**

As Verdejo (2006) explains, participant observation is used to get to know the reality of the community and to build up a certain amount of trust in order to share time with the community members. In short, participant observation no longer proposes

than to "walk around with my eyes open" and take advantage of the opportunity to share some moments of everyday life with the farmers.

- **Production and commercialisation flowcharts**

To create the production and marketing flowcharts, the Itaoca women's group met and collected information on the inputs used, the plants grown, the products manufactured and marketed or used. The materials used to carry out this activity were a large piece of paper, flipcharts, cardboard and paintbrushes.

- **Economic analysis**

According to Garcia Filho (1999), the economic analysis of production systems makes it possible to: i) assess the potential for capitalisation or, conversely, decapitalisation of each category of producer; ii) study in greater depth the social relations that characterise each type of production unit and the agricultural system as a whole; iii) understand the economic foundations of the associations of activities and agricultural practices adopted by producers. This assessment can be made from two points of view:

- That of the producer, who is concerned about the agricultural income that the production system can offer him;
- That of society, which is interested in the amount of new wealth generated by the production system.

In this case, the point of view being considered is that of the producer (Movimento Mulheres da Itaoca), who wants to know whether the medicinal plant business could generate profits or losses.

From this analysis it is possible to assess the potential for capitalisation or, conversely, decapitalisation of the production system. Based on Garcia

Filho (1999) explained what an economic analysis is to the Women's Movement of Itaoca. In order to give the Movement autonomy to manage its own business.

This analysis takes into account the added value of labour productivity, agricultural income and monetary income, as shown below:

1. **Added value**: Understanding the concept of added value allows us to analyse the idea of exceeding expectations in relation to the benefits provided by a good, i.e. actions added to a good (item, service, product), so that a comparison can be made between the previous process and the current one.

According to Garcia Filho (1999), by adding labour to the inputs and fixed capital at his disposal, the producer generates new wealth, adding value to these goods. The value added (VA) of the production system is equal to the value of what was produced minus the value of what was consumed:

VA = PB - CI - D, where:

GDP: Gross Product

CI: Internal consumption

D: Depreciation

In the system, it is also important to calculate labour productivity. This is obtained from: **Labour productivity=** VA/T, where:

VA: Value added

T: Number of workers employed in the system (including family members).

2. **Agricultural income:** Not all of the value added by the producer is intended to pay him or his family; he must rent land and pay a fee to the owner of the property, so if necessary, it must be included in the calculation of the producer's final income. The same goes for interest (a fee paid by the producer in exchange for advances from banks) and taxes (a fee paid to the state for the right to produce). He may also need salaried labour (permanent or temporary), in which case he pays wages. As a result, the value added in the establishment often has to be "shared" between the wage labourers, the owners of the land, the bank or the state. Thus, the part of the added value that remains with the producer after this distribution constitutes agricultural income (GARCIA FILHO, 1999).

We obtain the **Agricultural Income: RA** = PB - CI - D + Sub - S - I - J - RT, where:

S: Wages,

I: Taxes,

J: Interest,

RT: Land rent (leases) and

Sub: Subsidies.

Monetary income can also be obtained when disposable income is low or when producers have to make large cash payments (credit repayments, high-cost crops, etc.) (GARCIA FILHO, 1999).

RM = RA - self-consumption.

3.4 Production systems and analysing agroecosystems

According to Dufumier (1996), on the scale of an agricultural establishment, the production system can be defined as a combination (in time and space) of the resources available for obtaining plant and animal production. It can also be conceived as a more or less coherent combination of various production subsystems. In the case of the Itaoca Women's Movement's vegetable garden, it is a question of analysing the relationships of synergy or complementarity regarding the use of resources. In this way, analysing a production system on the scale of agricultural establishments is not just about studying each of its constituent elements, but above all about carefully examining the interactions and interferences between them.

Analysing agro-ecosystems is becoming important and highly relevant due to the transformations that the environment has undergone, as well as the need to analyse the actions that man has brought to bear on the environment.

In this way, the MESMIS methodology (Marco para la Evaluación de Sistemas de Manejo incorporando Indicadores de Sustentabilidad) helps to assess the sustainability of natural resource management systems, with an emphasis on the context of peasant and local producers, based on the storyline for the community, taking into account the participation or not of the subjects involved; provides critical reflection in order to improve the chances of success of alternative management systems and a holistic vision to understand the limitations and possibilities of sustainability management systems that arise from the intersection of environmental processes in the social and economic spheres (MASERA et al. 2000).

3.4.1 Subjects' perception of the work carried out

In order to understand the perception of the movement, we took as a basis their life stories, the relationship they have with the countryside, the importance of working in the garden and the effect of this activity on their lives and those of their families. The information was collected in the space itself, respecting the subjects' working hours. People who work in the garden on a daily basis were approached. All the data was collected using a voice recorder and recorded in a field diary. The data was analysed according to Garcia Filho (1999), by studying documents relating to history, rural development and agricultural practices.

3.1.2 Characterisation of the plant biodiversity cultivated in the garden of the

Itaoca Women's Movement

The existing flora was observed on a walk around the site and characterised in terms of scientific name and therapeutic indication based on Matos (2002). However, it cannot be said that the species found at the site are the same as those catalogued by this author, since the species were not identified taxonomically.

CHAPTER 4

CONTEXT AND RELEVANT CONCEPTS

4.1 What are medicinal plants?

Over the course of evolution, man has learnt to select plants for his diet and to relieve his ailments and illnesses. The result of this process is that many peoples have come to master the use of medicinal plants and herbs (FERREIRA & ANGELO 2010).

According to Jorge (2009), a plant is considered medicinal because it contains substances that have a pharmacological action (the action of the plant's chemical components on the body). These substances are called active principles, i.e. all the active principles form a phytocomplex that interacts with each other and with other apparently inactive molecules. This phytocomplex represents the integral pharmacological unit of the medicinal plant and this constitutes a therapeutic unit.

4.2 History of medicinal plants

Since time immemorial, humans have sought resources in nature to improve their own living conditions and increase their chances of survival (LORENZI & MATOS 2008).

The use of medicinal plants to restore health has evolved over time from the simplest forms of local treatment, probably used by cavemen, to the technologically sophisticated forms of industrial manufacture used by modern man (MATOS, 2002).

According to Jorge (2009), in the year 3000 BC there is information that China was dedicated to the cultivation of medicinal plants. Emperor Sheng-Nung used a series of plants on his own body to find out what effect they had. Amongst many, he emphasised the use of ginseng root, proclaiming it to be the most fabulous of herbs and that it favoured longevity, writing a treatise called PEN TSAO, a true pharmacopoeia that encompassed all the knowledge related to the use of plants as medicines. The same author also mentions that the first known Egyptian doctor was Imhotep (2980 to 2900 BC), a great healer who used medicinal herbs in his magical preparations.

In Brazil, the use of plants as medicine was influenced by indigenous, African and European cultures. Among the Indians, the shaman or sorcerer used narcotic plants to dream of the spirit that would reveal the herb or the way to cure the sick person, and also observed animals that sought out certain plants when ill (JORGE, 2009).

Lorenzi and Matos (2008) also stated that the effect caused by some plants inadvertently ingested also contributed to elevating plants to the category of divine entities and that plants with hallucinogenic properties were quickly included in religious rituals.

Another time when medicinal plants were widely used, according to Jorge (2009), were the years during the World Wars, when interest in medicinal plants returned due to the need to obtain effective remedies for multiple ailments, given that the entire economy of the countries involved in the war was earmarked for the production of war material.

Plants have always been linked to man and will always be used by him, both to cure ailments and for other multiple uses (JORGE, 2009).

4.3 The role of healers

The historical figure of the healer, popularly known at street markets as a herbalist, herbalist or root moulder, has stood the test of time and the innovations of scientific medicine, enduring to the present day and being much sought after by the population to buy medicinal herbs, parts of them or products already made and aimed at treating health problems (FRANÇA et al. 2008).

Raizeiros play an important role in the trade of medicinal plants and products in various municipalities in Brazil, especially in the Northeast (FREITAS et al. 2012). Most of the time, their knowledge is acquired from the family or through personal interest, and they seek out training/workshops on the subject. They are fundamental agents in the maintenance, transmission and dissemination of popular knowledge about plants and their respective uses. This knowledge needs to be rescued, valued and preserved.

4.4 Development of phytotherapy in the Unified Health System

Phytotherapy is a "therapy characterised by the use of medicinal plants in their different pharmaceutical forms, without the use of isolated active substances, even if they are of plant origin" (BRASIL, 2006).

The use of nature for therapeutic purposes is as old as human civilisation and, for a long time, mineral, plant and animal products have been fundamental to healthcare. Historically, medicinal plants have been important as herbal medicines and in the discovery of new drugs, with the plant kingdom making the greatest contribution to medicines (BRASIL, 2012).

The use of medicinal plants in primary health care programmes can be a very useful form of therapeutic alternative, due to their effectiveness combined with a low operational cost, given the relative ease with which plants can be acquired and the cultural compatibility of the programme with the population served. Adopting this resource is especially useful in communities where pharmaceutical medical assistance has proved difficult

(MATOS, 2002). In this sense, greater support is needed from both the medical and government sectors, the latter being more responsible for setting in motion laws and guidelines aimed at perpetuating the use of medicinal plants.

Integrative and Complementary Practices fall under what the World Health Organisation (WHO) calls traditional medicine and complementary and alternative medicine (TM/CAM) and, on this subject, the WHO recommends that its member states draw up national policies aimed at integrating/inserting TM/CAM into official health systems, with a focus on Primary Health Care (PHC) (BRASIL, 2012).

In Brazil, in line with the WHO's recommendations, the National Policy for Integrative and Complementary Practices in the SUS (PNPIC) was approved at 2006, covering institutional guidelines and responsibilities for implementing/adapting medicinal plant and phytotherapy actions and services in the Unified Health System (SUS) (BRASIL, 2012).

Furthermore, according to Brasil (2012), the approval of the PNPIC triggered the development of policies, programmes and projects at all levels of government to institutionalise these practices in the SUS. At the federal level, the National Policy for Medicinal Plants and Phytotherapics stands out, also approved in 2006 by presidential decree, with guidelines and actions for the entire production chain of medicinal plants and phytotherapics.

Medicinal plants and their derivatives are among the main therapeutic resources of TM/CM and have long been used by the Brazilian population in their health care, whether in Traditional/Popular Medicine or in public phytotherapy programmes in the SUS, some of which have been in existence for more than 20 years. Among the Integrative and Complementary Practices in the SUS, medicinal plants and phytotherapy are the most present in the system, according to a diagnosis by the Ministry of Health, and most of the experiences take place in PHC (BRASIL, 2012).

According to Ferreira (2010), the pharmaceutical industry has benefited and continues to benefit from popular knowledge about the medicinal use of plants.

Actions to implement the guidelines of these national policies seek to expand the supply of services and products related to herbal medicine in the SUS, in a safe and rational way, by qualified health professionals, considering the subject in their singularity and socio-cultural insertion, promoting comprehensive care (BRASIL, 2012).

4.5 Living pharmacies and their sustainability

The Living Pharmacies Project was created at the Federal University of Ceará as a means of developing social pharmaceutical assistance for organised government or private communities, the implementation of which requires interactive collaboration between doctors, pharmacists and agronomists. With the name Farmácias Vivas (Living Pharmacies), the aim is to distinguish the type of garden where scientifically validated medicinal plants are grown from those that work with empirical plants that are still subject to scientific validation (MATOS, 2002).

Matos (2002) states that the project was organised under the influence of the World Health Organisation's recommendations on the use of medicinal plants in primary health care programmes, and the success of the Medicinal Plants Research Programme (PPPM) of the Ministry of Health's Medicines Centre (CEME). Its aim is to offer non-profit phytotherapeutic pharmaceutical assistance to communities where public health programmes are lacking, promoting the correct use of locally or regionally occurring plants with scientifically proven therapeutic activity.

4.6 Commercialisation of herbal products

The use of medicinal plants to treat illnesses is an ancient human practice. In recent years, there has been growing interest in the knowledge, use and commercialisation of medicinal plants and herbal products in Brazil and around the world, which has led to a major expansion of this market (FREITAS et al., 2012).

According to Junior et al. (2005), various traditional clinical procedures using medicinal plants have been recorded over time. Despite the great evolution of allopathic medicine since the second half of the 20th century, there are basic obstacles to its use by underprivileged populations, ranging from access to hospital care centres to obtaining tests and medicines. These reasons, combined with the ease with which medicinal plants can be obtained and their long tradition of use, contribute to their utilisation by populations in developing countries.

To this day, some peoples still make conscious use of traditional herbal medicines related to knowledge and practices that have been acquired over the centuries. However, it should be emphasised that the use of this traditional medicine is often due to a lack of access to the medicine, and it is in this scenario that profiteers appear who sell fake and miraculous herbal medicines (FERREIRA and ANGELO 2010).

According to Junior et al. (2005), a large part of the marketing of medicinal plants takes place in pharmacies and natural products shops, where plant preparations are sold with industrialised labelling. In general, these preparations do not have a quality certificate and are produced from cultivated plants, which mischaracterises traditional medicine, which almost always uses plants from the native flora.

In this sense, Azevedo and Moura (2010) state that growing medicinal plants with a view to commercialisation requires planning in order to keep production constant and of good quality.

4.7 Urban agriculture

According to Mougeot (2000, p.7), urban agriculture is seen as agriculture that is located within (intra-urban) or on the periphery (peri-urban) of a town, city or metropolis and cultivates or raises, processes and distributes a variety of food and non-food products, (re)utilising to a large extent human and material resources, products and services that are found around that area, and in turn supplies human and material resources, products and services to a large extent to that same urban area.

According to Pinto Cibb and Yves Cribb (2009), urban food production is practised by a large proportion of the urban population in developing countries, and comes in various forms. It therefore refers not only to the harvesting of vegetables and the cultivation of fruit trees in cities, but also involves different types of cultivation such as medicinal, aromatic and ornamental plants. In this sense, we understand the activity developed by the Women's Movement of Itaoca, in which they practice the cultivation and commercialisation of medicinal plants, in an urban space located on land vacated by the City Council.

However, whatever interpretation is given to the concept, in many different countries, urban agriculture is considered a very broad term, as it includes everything from animal husbandry and plant production to activities involving the transformation and re-utilisation of solid and liquid waste (waste recycling, compost production and water re-use) (CITTADINI et al, 2002).

The existence of unoccupied spaces, such as vacant lots, needs to be cleaned up because they represent a means of accumulating rubbish and, consequently, favour the appearance of vectors such as rats, mosquitoes, flies, cockroaches, among other insects that are dangerous to the health of the population. Thus, encouraging investment in idle public or private areas improves the conditions of underused backyards, guarantees an improvement in the population's quality of life, both from a nutritional point of view - aiming for a better diet without the use of chemical inputs - and from the use of available labour, as well as providing considerably higher quality to the local environment through green areas, especially family community gardens, the use of ornamental and medicinal plants and urban afforestation with trees of various species (PINTO CRIBB and YVES CRIBB, 2009).

According to the same authors Pinto Cribb and Yves Cribb (2009). The experiences of agricultural activities in cities in various countries and in Brazil have shown that urban agriculture is an efficient option for community development and economic growth for poor families in urban areas. The benefits of urban agriculture are manifold and notably involve issues such as health, nutrition, combating poverty, sanitation, valuing local culture and, especially, environmental education, and can contribute greatly to the sustainable development of cities

CHAPTER 5

THE ITAOCA WOMEN'S MOVEMENT

5.1 Profile of the people who make up the Itaoca Women's Movement

The five people who make up the Movement range in age from 50 to 75. Mrs Silvino, from Limoeiro do Norte, has three children. She is one of the most active people in the group and sees the activity as a natural therapy. Mrs dos Santos, from Quixeramobim, is the mother of six children. Mrs Maria Estela, a native of Itapebussu and mother of four, has always loved the countryside. The oldest of the group, Mr José, was born in São Benedito and because he didn't like the countryside, he came to live in Fortaleza with his wife and children. Mr Mendonça, a native of Baturité, with three children and originally from the countryside, works as a volunteer and describes himself as an all-hours helper.

The rural origin of the majority of the group apparently explains why they enjoy working in the garden. On the other hand, the advanced age group of the Movement's members suggests that knowledge about medicinal plants is preserved and passed on by older people, in agreement with Alves et al. (2008), in the sense that older people tend to have greater knowledge about the therapeutic properties of plants and animals. The advanced age of these people means that they are limited in carrying out activities that require greater physical effort in the garden, which puts them in a situation where they have to rely on outside help to carry out heavier routine work, particularly the production of medicinal plants.

5.2 History of the Itaoca Women's Movement

It began in 2004 when an organised group from the community called the Movement of Popular Councils (MCP), a social movement that set about resolving problems in the neighbourhood, such as: violence; drugs; alcohol; lack of opportunities, among others. Formed essentially by women and only two men, the group was initially renamed Movimento da Itaoca (Itaoca Movement), then became Mulheres em Movimento da Itaoca (Itaoca Women in Movement), Flores da Itaoca (Itaoca Flowers), Mulheres da Itaoca (Itaoca Women) until it reached its current name:

Movimento Mulheres da Itaoca (Itaoca Women's Movement). This last name prevailed due to a typing error in the labelling.

They then started working in the field of education, and began to work more strongly on health issues, since the neglect of public health centres affected the poorest people. The land on which they started growing medicinal plants belongs to the City Hall and houses an Integrated Education and Health Centre (CIES). There used to be a Farmácia Viva (Living Pharmacy) in the outbuilding of the health centre, which was run by a team of people who didn't seem to be trained or to enjoy growing medicinal plants; in the eyes of the current management, the former employees only took care of it out of obligation. We don't know if it was because of this lack of training or even neglect that the living pharmacy failed to thrive. In terms of structure, there was a kitchen for manipulation, washbasins, a storage and drying room, laboratories and well-structured beds.

It was in operation for about two or three years, with a physical structure in good condition, an equipped laboratory and large beds for growing medicinal herbs. The space spent some time inactive until the organised group, made up mostly of women, together with the MCP, began to pursue a project to revitalise the Farmácias Vivas.

It's worth noting that when the women's movement arrived at the site, they found a lot of abandoned land, thrown away medicines, untended laboratories and flowerbeds, in other words, a space full of rubbish and weeds. Due to the conditions it was in, this contributed to facilitating the entry of bandits and crime into the area. The arrival of the movement gave a new meaning both to the space and to the lives of each member of the team, given that the women who took part in the group were those with problematic situations at home, with drug and alcohol problems and, above all, on low incomes. So they found an alternative there to overcome these problems, being a

a question of survival, as well as a new meaning to their lives. As Magilce Diniz, President of the Joazeiro Institute, explains, the participants were selected with the help of the community itself. "We approached the neighbourhood women's council and asked them to nominate needy women with low levels of education who were interested in taking part" (Universidade Pública, 2006).

The Joazeiro Institute for Sustainable Development is a non-profit civil society institution founded in October 2001 in the state of Ceará. Its strategy was to set up Reference Centres aimed at strengthening community spaces through the implementation of community urban gardens, training in the preparation of food with full use of nutrients, raising awareness and disseminating the use of solar energy and the rationalisation of water resources (Instituto Joazeiro, 2015).

The Institute was sought out because it is at the forefront of projects aimed at sustainable development of social, environmental and technological interest. At the time, the Institute had four neighbourhoods to start the project, but as the women's movement was strong and organised, this was the main key to implementing the medicinal plant project in Itaoca.

Through the "Reference Centre for Sustainable Urban Agriculture" (CRAUS) project, the Institute, together with the women's movement, began its activities in February 2006, with the support of the Secretariat for Economic Development, Fortaleza City Hall and the Federal University of Ceará, the latter contributing to the training of the agents involved.

As there was already a physical structure in place for a new medicinal plant cultivation, the beds, the laboratory (plant drying and storage greenhouses), the washbasins and part of the kitchen were utilised. The kitchen was extended to accommodate the handling practices for preparing home-made medicines and food.

At first, the movement had the active participation of 30 women and two men (multiplier agents), who underwent training and improved what they already knew about medicinal plants.

Initially, the team took turns every two days of the week to carry out the activities, working from 7am to 10am, receiving a hot meal to take home. The project made it possible to receive a grant of 30.00 reais for nine months and during this time the activities in the vegetable garden were carried out without any major problems, but as the grant came to an end, people started leaving and looking for other demands, from what was initially thirty-two people, today there are only five. There are three women and two men who really identify with the cause and do it because they like it.

Most of the activities in the first stage of the project, training, took place in the neighbourhood itself, using the CIES Professor Valdivino de Carvalho space, where the vegetable garden was set up and used for growing and harvesting plants. What was once rubble and scrubland was recovered and adapted for the work.

But the theoretical and practical classes also took place in other places, such as areas provided by the Centre for Agricultural Sciences (CCA), which supported the initiative by providing the infrastructure for field work (Public University, 2006).

The group that received the training was accompanied by students or professionals from Electrical Engineering, Agronomy, Home Economics and Pharmacy. When it came to preparing herbal medicines, the group was accompanied by an agronomist and a pharmacist who are partners of the Joazeiro Institute. In this sense, the involvement of other professionals is due to the fact that the training was not limited to the handling of medicinal plants, but also involved guidance on food handling, preparation and cooking (Universidade Pública, 2006).

As well as training in the use, handling and commercialisation of medicinal herbs, the group also received an extension project from the UFC, in which they would use solar energy in the water pumping system for irrigation of the medicinal herbs, which is part of the CRAUS project. The installation of the photovoltaic system made it possible to pump the water used to irrigate the vegetable garden. Pumping water using solar energy is viable because it is a non-polluting, sustainable source that is available all year round. And the process of irrigating the cultivated plants finds suitable conditions in the locality because the groundwater is easily accessible (Public University, 2006).

As a result of the group's initial organisation with the MCP, the idea of an alternative source of income and at the same time an escape from the problems in which they found themselves was taken forward. The Itaoca Women's Movement began to gain strength with the Joazeiro Institute, which revived the use, handling and commercialisation of medicinal herbs. The group began to receive a variety of training programmes. These included training to become better citizens, including political and citizenship training, notions of entrepreneurship, co-operativism and strategies for disseminating the knowledge acquired to the community (Chart 1).

All the products obtained from handling the herbs would, with experience, be used for their own consumption or by the residents who come to the health centre or neighbourhood markets (Public University, 2006).

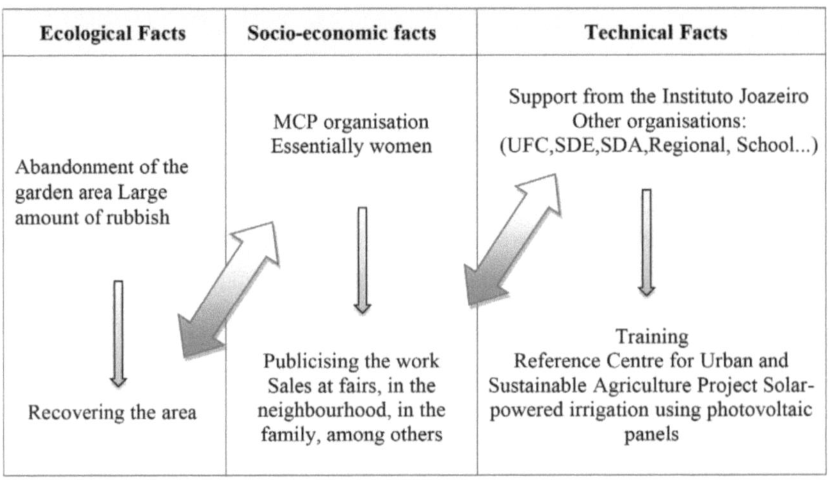

Figure 3: Representation of the evolution of key events in the Women's Movement of Itaoca, Fortaleza, Ceará, 2015.

Source: Research data.

When they arrived at the site, they found a lot of weeds and rubbish lying around, medicines piled up and the whole structure abandoned. The women reclaimed the space, cleaned it up, expanded the kitchen, an area for washing the plants, overhauled the compost in the beds and gave the place a new lease of life.

5.3 The social relations of the Itaoca Women's Movement

The social relationships developed by the Movement are illustrated by the flower-shaped model, where the petals are closest to the centre, indicating the most active relationships, and the sepals, furthest away, indicate the least active relationships today (Figure 4).

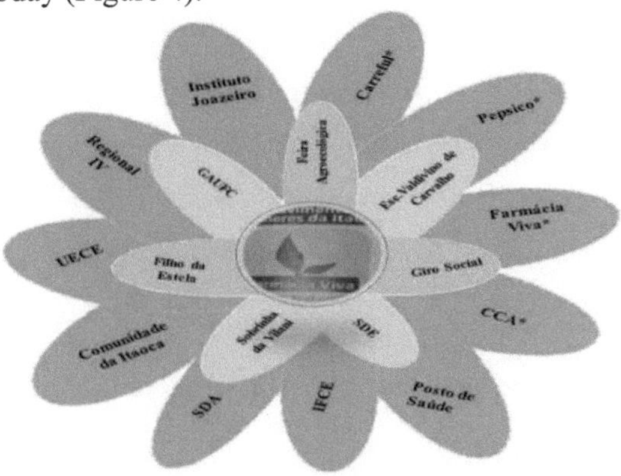

Figure 4: Representation of the social relations of the Itaoca Women's Movement, Fortaleza, Ceará, 2015.

Source: Research data.

Starting with the sepals, there is the Joazeiro Institute, the REGIONAL SECRETARIAT IV, the Federal Institute of Ceará (IFCE), the State University of Ceará (UECE), the Agrarian Development Secretariat (SDA), the Valdivino de Carvalho Health Centre and the Community of Itaoca. This group of organisations was more important in the past, as follows.

The Joazeiro Institute, with its social development project, provided workshops and training for the group. It brought with it various partnerships. However, with the end of the project, these partnerships disappeared, as did the Institute. What remained for the Movement was the learning and the initial impetus they provided. It was in Regional IV that the Movement began its activities with medicinal herbs in 2004. They also had the collaboration of IFCE, when students from the Institute installed the irrigation system for the beds. UECE collaborated with a technician who was in charge of the university's living pharmacy. He provided training on medicinal plants to the group. The SDA donated a forage machine to the movement to help make the compost. Previously, the health centre passed on some of the natural products donated by the movement to the community. Few people in the Itaoca community help out in any way in terms of involvement in the women's vegetable garden. Few people are interested in volunteering, they only help out when they are paid a small salary. Some people buy the produce. At the start of the project, the community was more present.

Continuing with the petals, there is the Valdivino de Carvalho School, the Ceará Electricity Company (Coelce), the Secretariat of Economic Development (SDE), the Benfica Agroecological Fair, the Agroecological Group of the Federal University of Ceará (GAUFC) and family members of the group members. These close relationships make more sense in the present.

At Valdivino de Carvalho School, the Movement publicises its work through workshops, field days, lectures and training for the community, involving teachers, children and their parents. With Coelce, through the Giro Social project, the Movement takes its products once a month to be sold in a space provided within the institution's premises. From SDE, the group occasionally receives donations of equipment, inputs (seeds and fertilisers) and gardening tools. When courses are offered at the institution, the Itaoca Women's Movement is invited to take part. The Benfica Agroecological Fair, which takes place fortnightly in Fortaleza's Gentilândia Square, is a public space where the group exhibits its products. They are well received by a clientele that is looking for their well-being. The GAUFC, made up of students from the UFC, has been helping to organise the medicinal plant production space and doing agronomic work. There are also family members of some of the group's members who help with the creation and manufacture of labelling and publicity, mainly on the internet.

CHAPTER 6

THE PRODUCTION SYSTEM AND LABOUR RELATIONS IN THE GARDEN OF THE ITAOCA WOMEN'S MOVEMENT

6.1 The Itaoca Women's Movement's medicinal plant garden

The implementation of the Farmácia Viva in Itaoca was conceived by a group of women with the support of the Joazeiro Institute, through the project of community gardens in the neighbourhoods.

The Valdivino de Carvalho health centre, on whose land the activity takes place, belongs to Fortaleza City Hall (Figures 4 and 6).

The data on the size in m^2 is not known to the group, the site has solar energy, but it no longer works and some gardening equipment has been donated, such as the forage planter.

There are 12 beds, six with medicinal plants and six with vegetables. Apart from the tyres, in which medicinal plants are also grown. The seedlings were brought in at the start of the project by the Joazeiro Institute, and are currently propagated by cuttings. The seeds are mostly donated and are hardly ever bought on the market.

The water used to irrigate and process the medicinal plants comes both from a local cistern and from the Ceará Water and Sewage Company (CAGECE). The energy comes from COELCE. Both water and energy are paid for by the health centre.

Figura 5: Construction of the map of the Itaoca Women's Movement garden, made by Seu Mendonça, Fortaleza, Ceará, 2015.

Source: Research data.

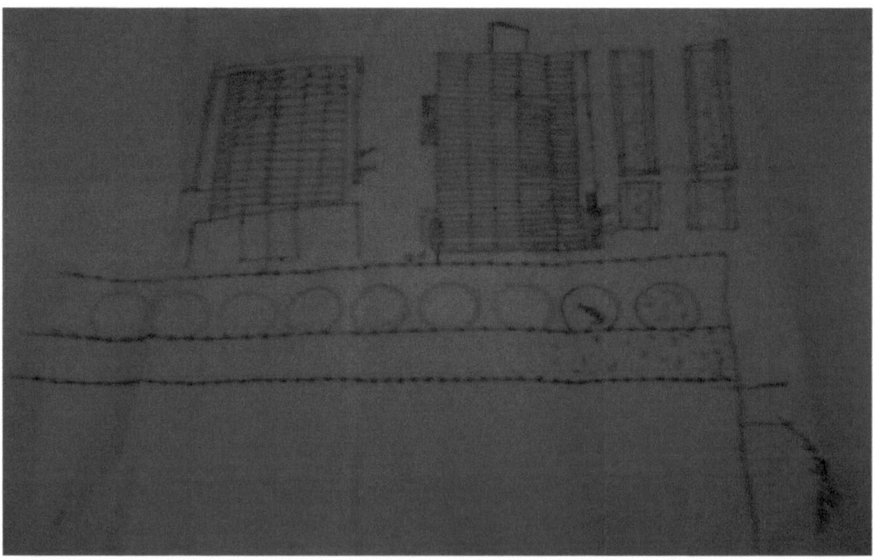

Figura 6: Map of the Itaoca Women's Movement garden, Fortaleza, Ceará, 2015.

Source: Research data.

The production area is well lit, which favours the development of the plants. The soil is sandy, considered by members of the Movement to be of inferior quality. However, according to Jorge (2009), species whose purpose is to extract roots, such as ginger, turmeric and saffron, prefer looser, sandier and less clayey soils. Species that produce a lot of leaf mass, such as mint, pennyroyal, lemongrass and lemon balm, prefer soils rich in organic matter.

The plant biodiversity cultivated is essentially made up of medicinal plants, fruit trees and vegetables (Table 2).

The spatial and temporal distribution of plant species depends on the inputs available on site, for example: medicinal plants are grown in beds, tyres and reused milk cartons, the latter of which are sold. The group carries out their activities according to the amount of labour they have, they don't follow fixed activities, so they plant/replant when they need to.

Figura 7: List of plant biodiversity cultivated in the garden of the Women's Movement of Itaoca, Fortaleza, Ceará, 2015.

	Medicinal plants grown in the garden	
Species	**Scientific name**	**Therapeutic indication**
Rosemary chilli	*Lippia sidoides* Chain.	antiseptic and antimicrobial
Chambá	*Justicia pectoralis var. stenoplrylla* Leon	asthma attacks, coughs, bronchitis and difficult breathing
Cashew	*Anacardium occidentale* L.	treatment of inflammation of the mouth and throat
Mastic	*Schinus terebinthifolius* Raddi	anti-inflammatory and healing
Good evening	*Catharanthus roseus*	used to cure cancers such as lymphomas
Babosa	*Atoe eraa* (L). *Burm.* F.	healing, antimicrobial
Holy grass	***Cymbopogon citratus* Stapf.**	digestion, nervousness, depression, agitation
Corama	***Kalanjoe brasiliensis* Camb**	headache, bronchitis cough
Malva santa	*Plettrantbus barbatus* Andr.	gastritis, dyspepsia, heartburn, gastric discomfort, hangover
Malvarisco	***Plectranthus amboinicus* (Lour.) Spreng.**	hoarseness, inflammation of the mouth and throat
Lavender	*Ocimumgratissimum* L.	expectorant and lung disinfectant
Creeping mint	*Mentha v villosa* Huds.	intestinal infestation, vaginal discharge
Mint vick	*Mentha arvensis* L.	malaise, stomach pains, headaches
Mentrasto	*ygeratum conyzoides* L.	analgesic, anti-inflammatory and febrifuge
Comfrey	*Symphytum qfficieale* L.	healing, fire burns, varicose ulcers
Cow's foot	*LauhiniaLonficata* Link	analgesic, laxative, diuretic
Safflower	*Curcuma longa* L.	treatment of constipation, cholesterol
Steel	*Malpighia glabra* L.	slows down ageing
Graviola	*Annona muricata* L.	antibacterial, anti-inflammatory, antifungal
Lemon	*Citrus limon*	infections, fevers, respiratory problems
Passion fruit	*Passiflora edulis* Sims	soothing and gentle sleep inducer
Citronella	*Cymbopogon nardus* (L) Rendle.	insect repellent and aromatherapy
Cider	*Lippia alba* (Mill)N. E. Brown	relieves colic and promotes the release of gas
Artemisia	*Artemisia vulgaris* L.	anaemia, colic, gastritis, nervousness
Losna	*Artemisia absinthium* L.	intestinal worms, colic, diarrhoea
Mulberry	*Rubus fruticosus*	mouth inflammations, sore throats, urinary tract infections
Brazilian Bilberry	*Plectranthus barbatus Andr.*	heartburn, stomach upset, hangover
Nirá	*Allium tuberosum*	are used to flavour salads and sauces.
Myrrh	*Commiphora myrrha* (T. Ness) Engl	strengthen gums, treat acne, canker sores

Source: Research data, supplemented by Matos (2002).

6.2 The system for producing, processing and marketing medicinal plants

The production, processing and marketing process of the garden's products is illustrated schematically in figure 6 and described below.

Figura 8: Flowchart of production, processing and commercialisation in the garden of the Women's Movement of Itaoca, Fortaleza, Ceará, 2015

Source: Research data

Soil preparation is carried out according to the availability of labour on site and consists of: turning over the soil along with the fertiliser and forming the beds and/or filling in beds made from cross-sectioned tyres.

The fertiliser used is a compost prepared on site from leaves and stems from tree and shrub pruning, shredded in a forage machine, weeding and food waste, enriched with earthworm humus, also produced locally.

The biodiversity cultivated in the garden is mainly renewed through seedlings and cuttings, with a much smaller number of plants propagated by seeds. The seedlings were initially donated by the Joazeiro Institute. They are currently multiplied on site through cuttings or division of rhizomes, generating a significant quantity of seedlings in the garden itself. However, seedlings are occasionally purchased from other producers or from the trade. The seedlings are prepared with the site's own substrate. Dry milk cartons are used to package the seedlings instead of black polythene bags, without harming the environment. The production of seedlings thus fulfils the internal needs of producing medicinal plants and generating income by selling seedlings at fairs. The seeds are bought in the shops and are coated with pesticides.

Planting is carried out at any time of the year, guided by the need, demand and availability of propagation material and labour. The criteria commonly used are based on daily observation of the plants in a position to provide cuttings or seeds and the condition of the seedlings for definitive planting.

Cultural practices are carried out manually, according to need, and include weeding, thinning, pruning, general cleaning of the area, top dressing and control of phytophages and phytopathogenic microorganisms. At the beginning of the year, these actions are more frequent due to the rainy season, which favours the appearance of phytophages and phytopathogenic microorganisms and the growth of spontaneous weeds, both in the beds and in their surroundings (Table 3, below). All the plant material removed goes into the compost heap, which is then returned to the beds in the form of fertiliser.

The most common phytophages are ants, mealybugs and aphids in warmer periods. In cold periods, snails appear. However, rarely is there any damage that warrants a control intervention. When necessary, homemade preparations are used. For mealybugs, powdered soap diluted in water is used. For aphids, use neem syrup. For snails, traps with beer or coarse salt are distributed. For ants, a solution of neem and chilli is prepared.

Irrigation of the beds is another practice influenced by seasonality, being more intensive in the dry season (Table 3).

Initially, leaves and vegetative material were harvested according to technical recommendations, i.e. at the most favourable times and days and according to the characteristics of each species. Nowadays, given the small number of people in the garden, it is usually carried out in the morning, because of the need to make the products and taking care not to harm the development of the plants.

Some of the harvested material goes for drying and/or processing, and some for seedling production. The fresh leaves destined for consumption, drying and/or processing undergo a health and integrity check.

Drying takes place in a natural environment, in the shade in a ventilated area, where the plants are protected from dust and free from insects and other animals. After drying, the material is stored in a preferably ventilated place.

The fresh or dried products are processed into tea, ointment, elixir, tincture, soap, licker and others every fortnight to ensure a constant supply of products. The processing operations are carried out at Mrs Vilani's house, due to the good space available.

The packaging, labelling and organisation of the products is carried out by the group in the garden space or in the larger one at Mrs Vilani's house.

The products are sold directly by the members of the group, without any intermediation, and there are some points where they have gained a loyal clientele for their products. Sales take place both at fairs and in the community of Itaoca, at different times, as follows: at the Bazar at Coelce and at the Benfica Agroecological Fair, sales take place every fortnight; at Coelce through the Giro social project, once a month; at fairs in different neighbourhoods; at the Valdivino de Carvalho school and at the homes of the group's members, in the latter two sales have no set period.

The group usually meets once a month to take stock of the activities and identify what needs to be improved in the garden.

6.3 Organising work in the Itaoca Women's Movement garden

Production, processing and marketing activities appear to have little dependence on seasonality. The effect of seasonality is more pronounced in the lower demand for irrigation and the higher incidence of spontaneous plants in the rainy season, requiring more labour in cultural practices, such as weeding, during this period (Table 3).

Table 1: Seasonality of activities in the garden of the Women's Movement of Itaoca, Fortaleza, Ceará, 2015.

Activity	Wet Period			Dry Period								
	J	F	M	A	M	J	J	A	S	O	N	D
Purchasing seeds	x	x	x	x	x	x	x	x	x	x	x	x
Seedling production	x	x	x	x	x	x	x	x	x	x	x	x
Cultural practices	↑	↑	↑	x	x	x	x	x	x	x	x	x
Planting	x	x	x	x	x	x	x	x	x	x	x	x

Irrigation	↓	↓	↓	x	x	x	x	x	x	x	x	x
Collecting leaves and plants	x	x	x	x	x	x	x	x	x	x	x	x
Product handling	x	x	x	x	x	x	x	x	x	x	x	x
Labelling	x	x	x	x	x	x	x	x	x	x	x	x
Commercialisation	x	x	x	x	x	x	x	x	x	x	x	x
Group meeting	x	x	x	x	x	x	x	x	x	x	x	x
Incidence of phytophages and diseases												
Stuffiness				x	x	x	x	x	x	x	x	x
Cochineal				x	x	x	x	x	x	x	x	x
Aphid				x	x	x	x	x	x	x	x	x
Snail	x	x	x									
Ants	x	x	x	x	x	x	x	x	x	x	x	x

Source: Research data.

Similar to seasonal activities, daily activities are practically constant throughout the days of the week (Chart 4). Both tables suggest an intensive routine and a consequent high demand for labour. This situation is common to all horticultural activities, as Lima and Gamarra-Rojas (submitted) have shown for other vegetable production systems.

Table 2: Weekly activities in the garden of the Women's Movement of Itaoca, Fortaleza-CE, 2015.

Activities	S	T	Q	Q	S	S	D
Irrigation	x	x	x	x	x	x	x
Cleaning the area	x	x	x	x	x		
Weeding	x	x	x	x	x		
Seedling production	x	x	x	x	x		
Sales	x	x	x	x	x	x	x

Source: Research data.

During the week, activities are carried out according to a work schedule defined by the group (Chart 5), although there are times when this schedule is changed and a member of the group takes over someone else's shift.

Table 3: Distribution of labour in the Itaoca Women's Movement, Fortaleza-Ce, 2015.

Group members	S	T	Q	Q	S	S	D
Mrs Vilani	x	x		x			
Mrs Esteia		x		x	x		
Mrs Lourdes		x	x	x			
Seu Zé (deceased)	x	x	x	x	x	x	X
Mr Mendonça	x	x	x	x	x	x	x

Source: Research data.

It can be seen that Mr Zé and Mr Mendonça are present every day, while the ladies are present three times a week, making sure that one of them is always present in the garden on weekdays. Mr Zé

takes care of the irrigation. The area is cleaned by Mr Mendonça. In his absence, weeding is done by someone from the community, who receives a small stipend for the work. The production of seedlings is carried out mainly by Mrs Estela. The products are processed by Mrs Vilani at her home. The product labels and the group's logo were created by the son of one of the Movement's participants. Mrs Lourdes and Mrs Estela are responsible for selling the products at fairs and other venues. Mrs Estela is in charge of recording and controlling the inputs, products and accounts.

It can be seen that the group has managed to mould or adapt its activity routines to its real working capacity, which is limited due to the advanced age of the Movement's members. However, this adaptation presents difficulties. Sometimes sacrificing technical aspects of production, sometimes processing and sometimes aspects of commercial activity, such as registering inputs and commercialised products.

6.4 The meaning of the garden for the members of the Itaoca Women's Movement

All the people in the garden have a connection to the countryside, as can be seen in their identification. They have different reasons for working in the garden or their motivation for this type of work.

Mrs Vilani explains that she does the work out of love, since the financial return isn't worth it. She also says that the day she doesn't go to the garden she feels ill. Her children admire what she does and collaborate indirectly by spreading the word about natural products on the internet or among friends. She makes a lot of use of natural remedies at home, and her neighbours and family, who know her work, look to her for an alternative to conventional remedies.

Mrs Lurdes says it's an old profession, as she used to work with her family in the fields since she was a child. Because she enjoys it so much, she spends more time in the garden than at home, which causes conflicts with her family. As for the benefits, she says that economically it's not so good, but it helps her well-being.

It's no different for Mrs Estela, who has always enjoyed working in the fields, helping her father in the fields since she was a child. She also says that her whole family uses the herbs. She also emphasises that working in the garden promotes a better quality of life.

Mr Zé spent some time working for companies and then retired. He started working in the vegetable garden, but, as he says, he doesn't completely enjoy it. Firstly, because he doesn't like the countryside; secondly, because the herbs don't grow well due to the sandy soil; and thirdly, because it doesn't provide a good income. He explains that the friendships he builds there are what motivate him to stay. The activity is a form of therapy for him, since he has health problems.

Mr Mendonça is very supportive of the garden's activities. He has a strong bond with it because he enjoys what he does and considers it therapy in difficult times. As he said, he suffers from depression. He sees the garden as a way of rescuing culture, as medicinal herbs are being lost over time.

Through the testimonies of each subject, it is possible to see the changes brought about by the simple fact of being in contact with nature. In this way, they relive the teachings passed down from their parents, putting their knowledge of medicinal plants into practice.

CHAPTER 7

ECONOMIC ANALYSIS OF THE DAITAOCA WOMEN'S MOVEMENT GARDEN

Through economic analysis, the members of the Itaoca Women's Movement will be able to manage their own business. They will be able to see the amount generated during the year by the medicinal plant garden in order to determine its productive potential, which is represented by the annual gross product (GDP) (total gross production of manipulated products + total gross production of seedlings produced). Of this amount, the self-consumption and commercialised value (represented by the sum of the products handled + seedlings produced).

Table 4: Breakdown of the annual gross product of the products handled in the garden of the Women's Movement of Itaoca, Fortaleza, Ceará, 2015.

Manipulated Products	Unit	Value (R$)
Compound lick	100 ml	5,00
Termite licker	100 ml	5,00
Nasal decongestant	30 g	6,00
Anti-inflammatory bottle	500 ml	20,00
Digestive elixir	60 ml	4,00
Digestive elixir	100 ml	7,00
Comfrey ointment	120 g	12,00
Ointment for cracked feet with sheep tallow	10 g	3,00
Ointment for cracked feet with sheep tallow	50 g	5,00
Rosemary tincture	60 ml	5,00
Citronella repellent	60 g	7,00
Mastic elixir	100 ml	8,00
Massage oil	100 ml	7,00
Medicinal oil	100 ml	7,00
Copaiba oil	100 ml	5,00
Aroeira intimate liquid soap	180 ml	15,00
Handmade mastic soap (bar)	50 g	5,00
Bath salt	50 g	3,00
Bath salt	100 g	5,00
Aloe vera shampoo	180 ml	15,00
Rosemary pepper antiseptic soap	100 ml	7,00
Chamomile liquid soap	100 ml	12,00
Passion fruit liquid soap	100 ml	12,00
Calendula intimate hygiene soap	100 ml	12,00
Rosemary chilli soap for intimate hygiene	100 ml	12,00
Comfrey cream	120 g	12,00
Inhalant for sinusitis	150 ml	10,00
Dried fennel plant	(50 g) bags	2,00
Rosemary mouthwash	60 ml	5,00
Arnica gel	120 g	12,00
Salt with herbs	145 g	4,00
Salt with herbs	350 g	12,00

Source: Research data.

Table 5: Breakdown of the annual gross product of the seedlings produced in the garden of the Women's Movement of Itaoca, Fortaleza, Ceará, 2015.

Seedlings produced	Unit	Value (R$)
Rosemary chilli	Seedlings	4,00
Chambá	Seedlings	4,00
Mastic	Seedlings	4,00
Good evening	Seedlings	4,00
Babosa	Seedlings	4,00
Capin santo	Seedlings	4,00
Corama	Seedlings	4,00
Holy Mallow	Seedlings	4,00
Malvarisco	Seedlings	4,00
Lavender	Seedlings	4,00
Creeping mint	Seedlings	4,00
Mint vick	Seedlings	4,00
Mentrasto	Seedlings	4,00
Comfrey	Seedlings	4,00
Cow's foot	Seedlings	4,00
Safflower	Seedlings	4,00
Acerola	Seedlings	4,00
Graviola	Seedlings	4,00
Lemon	Seedlings	4,00
Passion fruit	Seedlings	4,00
Citronella	Seedlings	4,00
Artemisia	Seedlings	4,00
Loana	Seedlings	4,00
Mulberry	Seedlings	4,00
Brazilian Bilberry	Seedlings	4,00
Nirá	Seedlings	4,00
Myrrh	Seedlings	4,00

Source: Research data.

Through a participatory conversation with the Movement's members, we analysed their internal consumption of the products they handle, noting that they make a lot of use of the compound licker, and we also heard that the biggest seller of these products is also the compound licker. This may be related to the fact that it is very easy for people to come down with unexpected colds, and it is indicated for coughs, hoarseness, sore throats, respiratory crises resulting from asthma and bronchitis.

With regard to the seedlings produced, they report that the biggest sales are of creeping mint, vick mint and citronella seedlings. This may be due to

This is related to the fact that the creeping mint seedling has great medicinal and social importance due to its action against intestinal microparasites, as well as in the treatment of amoeba, giardia and trichomonas (LORENZI and MATOS 2008). As for vick's mint, its widespread use and popularity can be related to its properties as a nasal decongestant and antipyretic, as well as its ability to relieve headaches (LORENZI and MATOS 2008). For citronella, its large commercialisation can be related to its insect repellent properties, especially mosquitoes. The internal consumption of seedlings by members is highest for malvariscus seedlings, followed by alfavaca.

For the production and handling of medicinal plants, some inputs are needed, such as: alcohol, moisturising base, labels, packaging, coriander seeds, labour, water, energy and fertiliser, the latter three being paid for by the Valdivino de Carvalho School, since a CIES operates on the land. So these inputs transformed throughout the process make up the garden's annual intermediate consumption (IC).

The annual depreciation (D) of fixed capital during the production process consists of the irrigation system and tools. The prices of these investments are obtained from the trade, as the Itaoca Women's Movement receives donations of these inputs.

In this way, the value added (VA) aims to express the new wealth of the Horto of the Women's Movement of Itaoca, which will emerge as the group adds labour and technology to the inputs and fixed capital it has at its disposal.

Looking at the system's total production, it can be seen that depreciation (D) is a minimal part of the value added (VA), which is a considerable part of the system's production.

Agricultural income (RA) plays a significant role in the total production of the garden, which will indicate the financial return it will bring to the group. In this way, sales of medicinal products and seedlings constitute a small income for the members. This, despite its small size within the garden, is essential, because even though it is an almost symbolic value, it is recognition that the group's work has been working, thus contributing to an increase in the production of natural remedies and valuing the performance of the Itaoca Women's Movement. It's worth remembering that in addition to sales, the members enjoy what they produce. For the members, self-consumption has a special meaning because they can follow the entire development of their work in the garden and have the opportunity to consume it. The members also reported that they often stop using conventional medicines (pharmacies) in favour of natural remedies (medicinal plants).

CHAPTER 8

FINAL CONSIDERATIONS

This study made it possible to visualise the organisation developed by the group to manage its activities, as well as helping to understand the socio-economic situation.

The subjects' perception of the group activities shows the degree of knowledge and affinity they have for spreading the use and reviving the tradition of medicinal plants.

The group carries out various activities in the garden, such as weeding, thinning, planting, replanting, drying, handling medicinal plants and marketing.

An important aspect of their work concerns their form of organisation and resistance to carrying out this activity, given that it does not bring a considerable economic return and that the use of industrial medicines is very frequent.

Analysing the Venn diagram shows that social relations with institutions that could provide greater backing and support for their work are weakened, in the case of Farmácia Viva, as well as the community in which they operate, because if they knew more about the work they do, they would be able to use and even help with the garden work more often. The tool shows that it is essential to find support to reinforce and disseminate the work they do, and that it is also interesting to look for projects to encourage more labour.

Analysing the group's history shows the degree of intimacy that the members already had from their outside experiences. This can be seen by analysing the identification of the individuals, showing their place of origin and their ways of working.

The economic analysis showed the possibilities for profits in the garden, as long as they manage the resources they use in the activity with exact control of what comes in and what goes out, making it easy to take stock of the economic return that could favour them.

As for the subjects' perception of the work they do, it can be seen that they all have a strong connection with the countryside and the activity provides well-being, although it doesn't bring a considerable financial return, the affective feeling for the garden is more rewarding.

It is hoped that this work will provide support for further research in this line of knowledge, valuing the production of urban gardens as well as the appreciation of popular culture.

REFERENCES

ALVES, R.R.N.; SILVA, C.C.; ALVES, H.N. Aspectos socioeconômicos do comercio de plantas e animais medicinais em áreas metropolitanas do Norte e Nordeste do Brasil. **Revista de Biologia e Ciências da Terra**, v.8, p.181-189, 2008.

APPOLINÁRIO, F. **Metodologia da Ciência:** filosofia e prática **da** pesquisa. São Paulo: Cengage Learning, 2009.

AZEVEDO, C.D; MOURA, M.A. **Cultivation of medicinal plants**: practical guide. Niterói: Rio Rural Programme, 2010.

BRAZIL. Ministry of Health. Health Care Secretariat. Primary Care Department. **National Policy for Integrative and Complementary Practices in the SUS - PNPIC-SUS** / Ministry of Health, Secretariat of Health Care, Department of Primary Care. Brasília: Ministry of Health, 2006, 92p. (Series B. Basic Health Texts).

BRAZIL. Ministry of Health. Health Care Secretariat. Primary Care Department. **Integrative and complementary practices:** medicinal plants and phytotherapy in Primary Care/Ministry of Health. Health Care Secretariat. Primary Care Department. Brasília: Ministry of Health, 2012.

CITTADINI, R.; GONZÁLEZ, N.; GONZÁLEZ, V.; CARROZZI, L.; GÉNOVA, F.; PORTA, J. La agricultura urbana como herramienta ante el proceso de marginación y exclusión en la ciudad de Mar Del Plata, Argentina. In: Proceedings of the VI Congress of the Latin American Association of Rural Sociology. UFRGS - Potro Alegre, November 2002. p. 472-478.

FRANÇA, I.S.X.; SOUZA, J.A.; BAPTISTA, R.S.; BRITTO, V.R.S. Popular medicine: benefits and harms of medicinal plants. **Revista Brasileira Enfermagem**. v.61, n.2, p.201-208, 2008.

FERREIRA, V. F.; PINTO, A. C. Phytotherapy in today's world. **Química Nova**, v. 33, n. 9, p. 1829-1829, 2010.

FREITAS, A.V.L. de; COELHO, M. F. B; AZEVEDO, R.A. de; MAIA, S.S.S. Os raizeiros e a comercialização de plantas medicinais em São Miguel, Rio Grande do Norte, Brazil.

Brazilian Journal of Biosciences. 2012. Available at: < http://www.ufrgs.br/seerbio/ojs/index.php/rbb/article/view/1863>. Accessed on: 20 January 2015.

GARCIA FILHO, D.P. **Guia metodológico de diagnóstico de sistemas agrários**. Brasília: INCRA/FAO, 1999. p. 42-44.

JOAZEIRO INSTITUTE, 2015. Available at: <http://www.institutojoazeiro.org.br/projetos> Accessed on: 22 June 2015.

JORGE, S.S.A. **Plantas medicinais**: coletânea de saberes, 2009. Available at: <http://www.fazendadocerrado.com.br/fotos_noticias/1280/livro.pdf>. Accessed on: 24 Feb. 2015.

JUNIOR, V. F.V; PINTO, A.C; MACIEL, M.A.M. Medicinal plants: a safe cure? **Chem. Nova**, v. 28, n.3, p.519-528, 2005.

LIMA, R.V.; GAMARRA-ROJAS, G. Peasants and the Mandalla in the Brazilian semi-arid region: reflections on sustainability from a case study with an agroecosystem approach. (submitted)

LORENZI, H.; MATOS, F.J.A. **Plantas medicinais no Brasil nativas e exóticas**. Nova Odessa: Instituto Plantarum, 2008. 512p.

MACIEL, M.A.M; PINTO, A.C; JUNIOR, V.F.V; GRYNBERG, N.F; ECHEVARRIA, A. Medicinal plants: the need for multidisciplinary studies. **Chem. Nova**, v.25, n.3, p.429-438, 2002.

MOUGEOT, L. J. A. Agricultura Urbana: concepto y definicón. La Revista Agricultura Urbana. Vol. 1, July 2000. Available at: < http:/www.ipes.org/aguila>. Accessed on: 13 March 2017.
DUFUMIER, M. **Les projets de développement agricole** - Manuel d'expertise. Paris: CTA-Karthala, 1996.

MASERA, O; ASTIER, M.; LÓPEZ-RIDAURA, S. **Sustainability and natural resource management.** The MESMIS evaluation framework. Mexico: Grupo Interdisciplinario de Tecnología Rural Apropiada; Instituto de Ecología; Mundi-Prensa, 2000. p.109.

MATOS, F.J.A. **Farmácias vivas**: a system for using medicinal plants designed for small communities. 4ª ed. Fortaleza: Editora UFC, 2002.

PINTO CRIBB, S. L. de S; YVES CRIBB. Urban agriculture: an alternative for alleviating hunger and for environmental education. **Sober** 47th Congress. 2009.

PUBLIC UNIVERSITY. Na Berlinda, Fortaleza, year 6, n. 34, Nov./Dec. 2006.

VERDEJO, M.E. **Participatory rural diagnosis:** practical DRP guide. Brasília: Ministry of Agrarian Development, 2006. p.6-37. Available at: < http://www.mda.gov.br/portal/saf/arquivos/view/ater/livros/Guia_DRP_Parte_1. pdf > Accessed on: 27 Nov. 2013.

ACKNOWLEDGEMENTS

My heartfelt thanks go to the Agroecology Group of the Federal University of Ceará (GAUFC), to the members of the group for having been part of this study and especially to my friend Ricardo Pereira, for his effort, dedication in every experience we shared and for having allowed me to meet incredible and courageous people like the Itaoca Women's Movement, where I was able to share knowledge and make great friendships that I treasure to this day.

My sincere thanks go to every member of the Itaoca Women's Movement (Mrs Vilani; Mrs Lurdes; Mrs Estela; Mr Mendonça) and the late Mr Zé, a great man who looked after the garden like no one else. I dedicate this work to my friend, who stopped his daily chores to help me while I was carrying out this study.

Thank you to everyone who somehow believes in the power of medicinal plants.

APPENDIX

1. Calculation of the annual gross product of the products handled in the garden of the Women's Movement of Itaoca, Fortaleza, Ceará, 2015.

Manipulated Products	Unit	Value (R$)
Compound lick	100 ml	5,00
Termite licker	100 ml	5,00
Nasal decongestant	30 g	6,00
Anti-inflammatory bottle	500 ml	20,00
Digestive elixir	60 ml	4,00
Digestive elixir	100 ml	7,00
Comfrey ointment	120 g	12,00
Ointment for cracked feet with sheep tallow	10 g	3,00
Ointment for cracked feet with sheep tallow	50 g	5,00
Rosemary tincture	60 ml	5,00
Citronella repellent	60 g	7,00
Mastic elixir	100 ml	8,00
Massage oil	100 ml	7,00
Medicinal oil	100 ml	7,00
Copaiba oil	100 ml	5,00
Aroeira intimate liquid soap	180 ml	15,00
Handmade mastic soap (bar)	50 g	5,00
Bath salt	50 g	3,00

Bath salt	100 g	5,00
Aloe vera shampoo	180 ml	15,00
Rosemary pepper antiseptic soap	100 ml	7,00
Chamomile liquid soap	100 ml	12,00
Passion fruit liquid soap	100 ml	12,00
Calendula intimate hygiene soap	100 ml	12,00
Rosemary chilli soap for intimate hygiene	100 ml	12,00
Comfrey cream	120 g	12,00
Inhalant for sinusitis	150 ml	10,00
Dried fennel plant	(50 g) bags	2,00
Rosemary mouthwash	60 ml	5,00
Arnica gel	120 g	12,00
Salt with herbs	145 g	4,00
Salt with herbs	350 g	12,00

Source: Research data.

2. Annual domestic consumption of products handled by members of the Women's Movement of Itaoca, Fortaleza, Ceará, 2015.

Products Manipulated	**Internal consumption of products by members**				
	Vilani	**Esteia**	**Zé**	**Lourdes**	**Subtotal**
Compound lick	120	6	100	60	286
Termite licker	12	6	50	0	68
Nasal decongestant	24	0	50	0	74
Anti-inflammatory bottle	48	2	10	1	61
Digestive elixir	120	120	10	3	253
Digestive elixir	48	0	0	3	51
Comfrey ointment	24	0	1	0	25
Ointment for cracked feet with sheep tallow	12	36	0	0	48
Ointment for cracked feet with sheep tallow	0	0	0	0	0
Rosemary tincture	36	72	10	2	120
Citronella repellent	48	0	0	0	48
Mastic elixir	12	0	0	0	12
Massage oil	12	0	0	0	12
Medicinal oil	12	0	5	0	17
Copaiba oil	0	0	0	0	0
Aroeira intimate liquid soap	24	12	2	0	38
Handmade mastic soap (bar)	24	0	2	0	26
Bath salt	0	0	0	0	0
Bath salt	0	0	0	0	0
Aloe vera shampoo	0	0	0	0	0
Rosemary pepper antiseptic soap	24	0	3	0	27
Chamomile liquid soap	0	0	0	0	0
Passion fruit liquid soap	12	0	0	0	12
Calendula intimate hygiene soap	0	0	0	0	0
Rosemary chilli soap for intimate hygiene	24	0	2	0	26
Comfrey cream	24	0	1	0	25
Inhalant for sinusitis	12	0	0	0	12
Dried fennel plant	24	0	3	0	27
Rosemary mouthwash	12	0	0	0	12
Arnica gel	12	3	0	0	15
Salt with herbs	36	0	0	1	37
Salt with herbs	0	0	0	0	0

Source: Research data.

3. Calculation of the annual gross product of the seedlings produced in the garden of the Women's Movement of Itaoca, Fortaleza, Ceará, 2015.

Seedlings produced	Unit	Value (R$)
Rosemary chilli	Seedlings	4,00
Chambá	Seedlings	4,00
Mastic	Seedlings	4,00
Good evening	Seedlings	4,00

Babosa	Seedlings	4,00
Capin santo	Seedlings	4,00
Corama	Seedlings	4,00
Holy Mallow	Seedlings	4,00
Malvarisco	Seedlings	4,00
Lavender	Seedlings	4,00
Creeping mint	Seedlings	4,00
Mint vick	Seedlings	4,00
Mentrasto	Seedlings	4,00
Comfrey	Seedlings	4,00
Cow's foot	Seedlings	4,00
Safflower	Seedlings	4,00
Acerola	Seedlings	4,00
Graviola	Seedlings	4,00
Lemon	Seedlings	4,00
Passion fruit	Seedlings	4,00
Citronella	Seedlings	4,00
Artemisia	Seedlings	4,00
Loana	Seedlings	4,00
Mulberry	Seedlings	4,00
Brazilian Bilberry	Seedlings	4,00
Nirá	Seedlings	4,00
Myrrh	Seedlings	4,00

Source: Research data.

4. Annual internal consumption of seedlings produced by members of the Itaoca Women's Movement garden, Fortaleza, Ceará, 2015.

Seedlings produced	Internal consumption of seedlings by members				
	Vilani	Esteia	Zé	Lourdes	Subtotal
Rosemary chilli	0	0	0	0	0
Chambá	72	0	2	0	74
cashew	0	0	0	0	0
Mastic	96	0	0	0	96
Good evening	0	0	0	0	0
Babosa	72	1	0	0	73
Capin santo	48	5	12	1	66
Corama	48	1	12	0	61
Holy Mallow	48	2	0	3	53
Malvarisco	120	2	1	10	133
Lavender	120	0	0	0	120
Creeping mint	72	0	1	1	74
Mint vick	72	0	1	1	74
Mentrasto	48	0	0	0	48
Comfrey	72	1	0	0	73
Cow's foot	60	0	0	0	60
Safflower	36	0	0	0	36
Acerola	0	1	0	1	2
Graviola	0	1	0	0	1
Lemon	0	2	0	0	2
Passion fruit	0	0	0	0	0
Citronella	48	1	0	0	49
Cider	72	1	12	1	86
Artemisia	36	1	0	0	37
Loana	12	0	0	0	12
Mulberry	12	2	0	0	14
Brazilian Bilberry	12	0	0	0	12
Nirá	24	0	0	0	24
Myrrh	0	0	0	0	0

Source: Research data.

5. Intermediate consumption in the garden of the Women's Movement of Itaoca, Fortaleza, Ceará, 2015.

Labour: R$50.00 per month for each member of the group, taking into account that there are 5 people working in the garden. Considering that the year has 12 months, the amount paid in labour per year will be R$3,000.00.

Monthly supplies and overheads	Price (R$)/month	Unit	Subtotal
70 % alcohol	14,00	1	15,00
Grain alcohol	24,00	1	25,00
Moisturising base	28,00	1	29,00
Transport (each journey)	60,00	1	61,00
Labels	4,50	1	5,50
Coriander seeds	12,00	1	13,00
Water	-	-	-
Energy	-	-	-
Fertiliser	-	-	-
Packaging (prices depend on size)			
Lick pots (centre) / recycled	35,00	100	35,00
Lick pots (centre) / not recycled	65,00	100	65,00
Liquid soap pots	2,00	30	60,00
Gel pots (small)	0,90	50	45,00
Gel pots (gr.)	1,20	50	60,00
Pots for comfrey ointment (small)	0,90	20	18,00
Pots for comfrey ointment (gr.)	1,20	20	24,00
Herb salt pots	0,20	100	20,00
Nasal decongestant pots	0,45	30	13,50
Dishwasher jars (small)	0,45	100	45,00
Garlic pots (gr.)	0,65	100	65,00
Repellent tubes	1,15	20	23,00
TOTAL ANNUAL OVERHEADS			**616,00**
TOTAL ANNUAL OVERHEADS + LABOUR			**3616,00**

Source: Research data.

6. Investments in the garden of the Itaoca Women's Movement, Fortaleza, Ceará, 2015.

Investment table

Products	Price (R$)	Qty. Used in the system	Qty Total - vi (R$)	Duration - n (years)	Depreciation - Vi/n
Santeno Tape (100m)	56,41	30	16,92	3	5,64
Hose Polyethylene 3/4 (lOOm)	104,31	30	31,29	3	10,43
Solar pump 8000 12 v (un.)	445,54	1	445,54	8	55,69
Hoe (un.)	18,90	6	113,40	2	56,70
Wheelbarrow (pcs.)	174,90	2	349,80	3	116,60
Shovel (pcs.)	19,76	3	59,28	2	29,64
Scissors (un.)	20,50	3	61,50	2	30,75
Freezer (pcs.)	1879,90	1	1879,90	8	234,99
Forage machine (un.)	613,70	1	613,70	8	76,71
TOTAL DEPRECIATION OF FIXED CAPITAL					**617,15**

Source: Research data.

ANNEXES

Annex 1: Physical structure of the Itaoca Women's Movement garden
Annex 2: Production beds in the Women's Movement garden

Annex 3: Growing medicinal plants in the garden of the Itaoca Women's Movement, Fortaleza, CE.
Annex 4: Cultivation of medicinal plants by the late Seu Zé, in the garden of the Itaoca Women's Movement,

Annex 5: Growing medicinal plants in the garden of the Itaoca Women's Movement, Fortaleza, CE.
Annex 6: Seedling production in the garden of the Itaoca Women's Movement, Fortaleza, CE.

Annex 7: Seedling production in the garden of the Itaoca Women's Movement, Fortaleza, CE.
Annex 8: Production of natural remedies in the garden of the Itaoca Women's Movement, Fortaleza, CE.

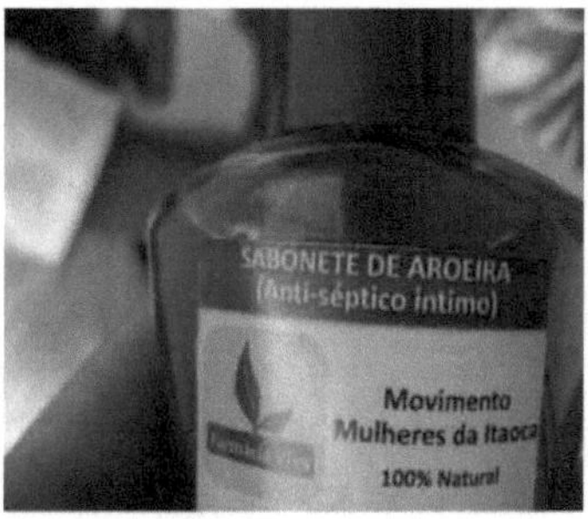

Annex 9: Production of natural remedies in the garden of the Itaoca Women's Movement, Fortaleza, CE.
Annex 10: Seal of the Itaoca Women's Movement, Fortaleza, CE.

ABOUT THE AUTHORS

Maria Edvania Neves Barros

She has a degree in Agricultural Engineering from the Federal University of Ceará (UFC) and is currently studying for a master's degree in Plant Science at the UFC with an emphasis on Biological Control/Acarology. She has experience in rural extension and medicinal plants.

Guillermo Gamarra-Rojas

She has a degree in Agronomy from the Federal University of Alagoas (1985), a master's degree in Agricultural Sciences from the School of Agronomy of the Federal University of Bahia (1994) and a PhD in Botany from the Federal Rural University of Pernambuco (2002). He works in the areas of Agroecology, Rural Extension, Development and Public Policies. He has experience in drawing up and coordinating research and development projects, extension and professional training at local, state and national level and in international cooperation. He is currently an adjunct professor at the UFC. He teaches on undergraduate courses in the Agricultural Sciences (Agronomy, Ecological Economics, Fisheries Engineering and Zootechnics). He is a supervisor and co-supervisor on postgraduate courses: Academic Master's Degree in Rural Economics (MAER), Postgraduate Programme in Development and the Environment (PRODEMA) and Interuniversity Master's Degree in Agroecology: an Approach to Rural Sustainability, International University of Andalusia, Baeza. In this work she participated in the conception and design of the research, the analysis and interpretation of the data and the writing and critical analysis of the text.

Alfredo Mendonça de Sousa

Graduated in Agronomy from the Federal University of Ceará. He has worked in the field of Agronomy, mainly in the areas of Irrigation and Agricultural Economics. He has teaching experience working as a monitor of the subject Rural Constructions and Rural Electrification in the Department of Agricultural Engineering - UFC and as a teacher/facilitator in secondary schools (pre-entry exams), working mainly in the subject of maths (volunteer teacher). He is currently studying for a master's degree in Agronomy - Soils and Plant Nutrition (with an emphasis on fertigation) at the Federal University of Ceará.

MIX
Papier aus verantwortungsvollen Quellen
Paper from responsible sources
FSC® C105338

Printed by Books on Demand GmbH, Norderstedt / Germany